Fruit
The Ripe Pick

Fruit
The Ripe Pick

Fruit Selection Made Easy!

T.M. Gorman

Honolulu, Hawaii

Book design by Pete Masterson, Aeonix Publishing Group, www.aeonix.com
Cover design by Robin Chang, e-mail: robinhi@stghi.com
Editing by Marilyn Weishaar, e-mail: mrsmom@nvc.net

Disclaimer

This book is intended to provide accurate information with regard to the subject matter covered. Every effort has been made to make this book as accurate and complete as possible. However, the author, publisher, and all entities accept no responsibility for inaccuracies or omissions. The author, publisher, and all entities specifically disclaim any liability, loss, or risk, whether personal, financial, or otherwise, that is incurred as a consequence, directly or indirectly, from the use and/or application of any of the contents contained in this book.

Published by
Vine Publishing
Post Office Box 17912
Honolulu, HI 96817-9998 U.S.A.
e-mail: vine@hawaii.rr.com

Publisher's Cataloging-in-Publication
(Provided by Quality Books, Inc.)

Gorman, T. M.
Fruit, the ripe pick : fruit selection made easy! /
T. M. Gorman. -- 1st ed.
p. cm.
LCCN: 00-135524
ISBN: 1-931141-20-7

1. Fruit. 2. Grocery shopping. I. Title

TX397.G67 2001 641.3'4
QBI00-847

Printed in the United States of America

Contents

Acknowledgments

This book is the sweet fruit of many people's labor. Without the love, support, encouragement, and guidance of my family, friends, and colleagues, I would never have been able to turn my dream into a reality.

I thank my husband, Richard, for sharing and encouraging my journey in developing this book, for the loving sacrifices he made, and for motivating me to pursue an idea that began as a seed in my mind years ago. I love you, and thank you for allowing me the time to create this project and pursue this aspiration.

I thank my Mom, Deborah, for giving me her unconditional love, and enduring with me through challenging times in my journey. I thank her for always being there while I laid the foundation for my future direction and goals, and for her humble, yet steadfast support. I thank my deceased father, Patrick, for his love, gentle spirit, kindness, and generosity that touched me, and all who knew him. To my loving sisters Patti, Roseline, Tahnee, and Tatum, thank you for your optimism and inspiring words of encouragement every step of the way. To my stepfather Vernon, I will always appreciate and respect you for your support and love for all of us.

I thank my graphic designer, Robin Chang, for his creative, artistic contribution and patience with me in designing the cover. I thank my editor, Marilyn Weishaar, for adding her talent, expertise, and professional editing skills. I thank Pete Masterson, my book designer, who added his perfecting and finishing touches.

Thanks to my dear friend, Lahilahi, for her loyalty, friendship, and her commitment in reading and giving honest feedback on early drafts. To my extended family, friends, and associates who offered valuable and encouraging feedback when I needed it most, I am truly grateful.

Special Acknowledgment

This book would never have been written without the brilliant creativity, unrelenting support, skillful wisdom, and contributions of my brother Herman. He spent untold hours researching, and perfecting my ideas. His inspiration kept this project alive when my spirits faltered, his tenacity kept me on track, and his enthusiasm guided this book from inception to completion. Thank you, brother, for putting your heart and soul into this project.

Last, but always first, I thank the Creator for the countless blessings He has given me, and for allowing me to share a small portion of these blessings with you.

Aloha,

T.M. Gorman

Introduction

Walk into any grocery store, supermarket, or open air market anywhere in the world and you are sure to find one thing in common, fruit of some kind. You'll also find something else at almost every fruit counter, people trying to decide which fruit to pick. Shoppers look for the brightest colors, check for obvious bruises, or squeeze to see if the fruit is hard or soft. If only it were that easy to select fruit that is sweet, juicy, and delicious!

Every fruit is unique, each with its own distinguishing characteristics and methods of selection. Many people, including myself, are often convinced that we have found just the right piece of fruit only to be disappointed when we bite into a dry, tart orange, or a mealy, flavorless peach.

This was my motivation for writing *Fruit – The Ripe Pick*. This book helps you pick delicious fruit the first time; a handy reference guide you can use again and again.

During my research, I learned much more about fruit than merely how to choose it. Each fruit has a distinct origin and history, trivia information, and unique uses. Adding some of those tidbits in a "Did You Know..." section for every fruit makes this book much more than just a selection guide. It entertains while it educates.

My hope is that you will enjoy and utilize this book wherever and whenever you buy fruit. May you get as much enjoyment out of your next piece of fruit as I got out of sharing this information with you, my readers.

Fruit
The Ripe Pick

Fruit Selection Made Easy!

Apple

Select apples that are firm and have a tight, smooth skin. They should be well colored for the variety and not yield to palm pressure. Brown spots, known as scald, usually will not affect freshness or flavor. Pay attention to firmness in larger apples; they tend to ripen faster than smaller ones.

Avoid apples that have soft spots, punctures, large discolored areas, or skin breaks and indentations.

Store in a plastic bag in the crisper compartment of your refrigerator where it's coolest and the humidity is high. Apples ripen quickly and should not be kept at room temperature for more than two days. Mushy pulp, especially in larger varieties, is an indication of prolonged or improper storage. Wash just before eating.

Popular Varieties: There are more than seven thousand varieties.

Best for eating raw: Fuji (juicy, crisp, sweet), Gala (rich, unique flavor), Golden Delicious (very sweet, tender flesh), Red Delicious (sweet, mildly tart), McIntosh (tangy, juicy).

Best for cooking: Granny Smith (tart, firm), Pippin (less tart, crisp), York Imperial (sweet, crunchy).

Best for baking: Cortland, Northern Spy, Rome Beauty.

Peak Season: Late Summer – Fall

Nutritional Content: 1 medium apple: 80 calories, 0.3 g fat, 22 g carbohydrates, 16 g sugar. Good source of vitamin C and fiber.

Did You Know...

- You can keep apple slices from browning by tossing them with lemon, orange, or grapefruit juice, or by dipping them in lightly salted water.
- Apples are members of the rose family.
- Apples float because twenty-five percent of their volume is air. They bruise easily because twenty-five percent of their weight is water.
- American John Chapman, better known as "Johnny Appleseed," spread apple growth by planting apple seeds and cuttings throughout the United States.
- During the 1920s and early '30s, Prohibition almost forced many apple varieties into extinction because they were used for making liquor. Apple cider, a substitute for liquor, became more popular than milk.
- Many biblical experts contend that a quince, a pear-like fruit, rather than an apple was the fruit eaten by Adam and Eve in the Garden of Eden.
- In ancient Greece, tossing an apple to a woman was a traditional way of proposing marriage; catching it meant acceptance.

Apricot

Select apricots that are plump, reasonably firm, fragrant, and have a consistent dark yellow or orange-yellow blush. The skin should be smooth and velvety. Depending on the variety, color can range from pale yellow to deep burnt-orange. Apricots should have a sweet, tart flavor. Always handle apricots gently; they bruise easily, which causes them to turn soft and lose their flavor.

Avoid apricots that are green tinged or bruised, and those that have white spots or appear shriveled or wilted. The latter indicates a lack of flavor and decay.

Store unripe apricots at room temperature until slightly soft, a sign of ripeness. Once ripe, refrigerate in a plastic bag up to one week. In order to speed ripening, store apricots in a paper bag at room temperature until ready to eat.

Popular Varieties: Blenheim (medium size, freestone, orange, sweet, aromatic), Harglow (medium size, bright orange blushed with red, firm, sweet), Tilton (medium to large, yellow-orange, rich flavor, eaten fresh or dried, best for canning), Tomcot (large, orange, firm, sweet). There are dozens of varieties; they range in blended colors from pale yellow to golden-orange.

Peak Season: June – July

Nutritional Content: 3 small fresh apricots: 60 calories, 0.4 g fat, 12 g carbohydrates, 11 g sugar. Rich supply of beta-carotene, high in vitamin A and vitamin C.

Did You Know...

- When dried, apricots are one of the best natural sources of vitamin A.
- Apricots are often used in making wine and brandy.
- Apricots are a primary food source of the Hunzas, a Himalayan tribe noted for its longevity; Hunzas often live past the century mark.
- Apricots were first grown in China more than two thousand years ago and were called "moons of the faithful" by Confucius, who ascribed his knowledge to its tree.
- Alexander the Great was the first to bring apricots to the Roman Empire, where they became known as "golden apples."

Asian Pear

Select Asian pears that are smooth and thin skinned. They should be extremely firm and heavy for their size; weight indicates juiciness. Depending on the variety, colors range from pale yellow and yellow-green to caramel and bronze. Asian pears are selected by smell; check for a fairly strong and sweet aroma. If the pears are cold, the smell will not be as intense.

Avoid Asian pears that have soft spots, cuts, scuffmarks, or bruises.

Store refrigerated up to one month.

Popular Varieties: Najesseki or 20th Century (most popular, medium size, golden-yellow skin, firm, juicy, white flesh), Shinko (medium to large, yellow-green skin with a russet blush, rich, sweet flavor), Large Korean (large, bronze-colored skin, juicy, sweet flavor).

Peak Season: July – October

Nutritional Content: 1 medium pear: 61 calories, 0.5 g fat, 29 g carbohydrates. Good source of fiber, source of vitamin A and vitamin C.

Did You Know…

- Asian pears make excellent pie filling and are great for teething babies because of the pear's firm texture and juicy flesh.
- Asian pears, also known as pear apples, suffer an identity crisis. Like apples they ripen on the tree, are firm and crisp, but taste and smell like pears.
- Asian pears are the oldest known cultivated pears; they were developed in Japan in 1898. Their original name is Najesseki (better known as 20th Century in the United States), currently the most popular.
- Chinese gold prospectors introduced Asian pears to the American West in the mid 1800s.

Avocado

Select avocados that are unblemished and heavy for their size. If you are not planning on using them immediately, look for firmness. When ripe, avocados will yield to gentle palm pressure. The "Hass" avocado turns dark green or black when ripe; the popular "Fuerte" retains its light green color.

Avoid avocados that have dark, sunken spots or are too soft, which are indications of being overripe. Once peeled, avocados darken quickly, and should be eaten immediately or sprinkled with citrus juice.

Store unripe avocados at room temperature for three to six days. In order to speed the ripening process, place them in a paper bag until ready to eat. Refrigerate ripe avocados up to three days. Do not refrigerate hard avocados; they will never ripen.

Popular Varieties: Hass (medium size, thick skinned, pebbly, purple-black skin, small seed) and Fuerte (large, smooth, dark green, thin skinned, medium size seed).

Peak Season: Hass: January – November, Fuerte: November – May

Nutritional Content: 1 medium avocado: 275 calories, 25 g fat, 15 g carbohydrates, 0 g sugar. Contains 17 vitamins and minerals, excellent source of potassium.

Did You Know...

- Avocados will ripen quicker if buried unwrapped in flour.
- Avocados contain more potassium than most other fruits and vegetables.
- Avocado leaf and seed extracts have been used as antibiotics and to treat diarrhea.
- Although avocados must reach full maturity before they are picked, once mature they can remain on the tree for many months.
- Avocados date back to 8,000 B.C., originating in Mexico and Central America. Their seeds have been uncovered with mummies in Peru dating back to 750 B.C.
- Until the 20th century, Aztecs believed avocados were a sexual stimulant and anyone concerned about their reputation did not eat them.

Banana

Select bananas that are firm, plump, and have a bright, even color. Tiny brown spots merely indicate ripeness. Depending on the variety, skin tone can vary in color from pale creamy yellow to russet red. The stems should be intact and the fruit free of skin splits; these are entry points for contamination.

Avoid bananas with soft, bruised spots, which indicate that they may be overripe. Once peeled, bananas darken quickly, and should be eaten immediately. A grayish cast may be a sign that they were stored at a cold temperature and may not ripen properly..

Store at room temperature until ripe. If refrigerating ripe bananas, the peel may turn brown, but the fruit will remain fresh up to two weeks. Peeled bananas can be frozen up to six months. Never refrigerate unripe bananas, it will stop the ripening process.

Popular Varieties: Mysore, Apple, Ice Cream, Red, Orinoco. There are three hundred varieties.

Peak Season: Available year-round

Nutritional Content: 1 medium banana: 150 calories, 0 g fat, 29 g carbohydrates, 21 g sugar. Source of vitamin C, fiber, and potassium.

Did You Know...

- Bananas are the world's largest herb.
- Sautéed green-tipped bananas are an excellent side dish for chicken, pork, or fish.
- Bananas are believed to have originated in the jungles of Malaysia because of the many varieties found there. They were once called "bannas," "ghanas," or "funanas" until Africans gave them their present name.
- Alexander the Great discovered bananas during his conquest of India in 327 B.C.
- Bananas were formally introduced to Americans at the 1876 Philadelphia Centennial Exhibition, where each one was wrapped in foil and sold for ten cents.

Blackberry

Select blackberries that are glossy, deep colored, and well formed. They should be fairly firm and without hulls. The white color found on blackberries is a natural protective coating.

Avoid blackberries that are dull colored, dry, soft, moldy, or bruised. If the hulls are attached, the berries may be tart and might never develop their full flavor. Check for seeping juices, which may be a sign that berries are overripe.

Store in the refrigerator up to two days. They should be covered, unwashed, and un-stemmed. Wash just before using. Blackberries are extremely perishable and should be handled with care. Damaged berries have less vitamin content.

Popular Varieties: Two main types: Albino and Dewberry. There are twenty-four species.

Peak Season: June – July

Nutritional Content: 1 cup blackberries: 60 calories, 0 g fat, 12 g carbohydrates, 11 g sugar. Good source of vitamin C and fiber.

Did You Know…

- Blackberries are members of the rose family and are native to North and South America, Asia, and Europe.
- Adding lemon juice to a blackberry pie will help blackberries retain their natural color.
- Willamette Valley in Oregon accounts for eighty-five percent of the domestic production.
- Blackberries have been used for two thousand years for medicinal purposes in Europe.
- Blackberry vines are used as natural barriers, much like a fence or wall.
- Although blackberries have always flourished in the wild, they have been domestically cultivated only since the turn of the 20th century; most notably by Fredrick Coville, a New Jersey botanist.

Blueberry

Select blueberries that are plump, firm, and uniform in size. They should have an even, silver-frosted indigo blue color. A whitish bloom indicates freshness. Shake the container; firm berries will move freely, softer berries will stick together.

Avoid blueberries that are soft, bruised, moldy, or seeping juices. If purchasing pre-packed berries, check bottom of basket for berry juice; this indicates berries are smashed, or possibly rotten.

Store in the refrigerator. Once chilled, they will retain quality for ten to fourteen days. Wash berries quickly and gently right before using. They can be frozen in a heavy-duty plastic bag up to one year. In order to avoid clumping, freeze berries on a cookie sheet prior to packaging.

Popular Varieties: Two types: Northern Highbush (supermarket variety, plump, blue-black color with a silver-white bloom, grown in northern and southern United States and British Columbia) and Wild Lowbush (rare, smaller, darker, chewy texture, more intense flavor, grown in Maine and Canada).

Peak Season: May – October

Nutritional Content: 1 cup blueberries: 100 calories, 1 g fat, 27 g carbohydrates, 11 g sugar. Good source of vitamin C and fiber.

Did You Know...

- You can prevent blueberries from bleeding in baked goods by adding frozen berries to batter rather than thawed berries, which may have broken skins.
- More than two hundred million pounds of blueberries, ninety percent of the world's supply, are harvested throughout North America each year.
- Blueberries are one of three fruits native to North America and are its second most favorite berry, making blueberry pie one of North America's most popular pies.
- Bears love blueberries and have been known to travel up to fifteen miles to eat them.
- Blueberries have been in existence for thousands of years and were once called "star berries" because of their star-shaped crowns.
- American Indians believed that the blueberry's five-point star blossoms were sent by the "Great Spirit" to lessen hunger in times of famine. They also used blueberry roots to make medicinal teas such as childbirth relaxants.

Boysenberry

Select boysenberries that are glossy, well formed, dry, plump, and reddish-purple in color.

Avoid boysenberries that are soft, bruised, or moldy. Always check berries or packaging for signs of seeping juices, which indicate the berries may be overripe.

Store unwashed berries in the refrigerator for two to three days, being sure to remove moldy or bruised berries before storing to prevent decay. Wash them quickly and gently just before using.

Peak Season: July – August

Popular Varieties: Hybrid

Nutritional Content: 1 cup boysenberries: 65 calories, 0 g fat, 16 g carbohydrates, 11 g sugar. Source of vitamin C and potassium.

Did You Know...

- Although the origin of boysenberries is unknown, it is assumed they are a cross between a blackberry and either a red raspberry or loganberry. They resemble elongated blackberries with a rich maroon-like color, but have a sharp, tart flavor.
- In the late 1920s, George Darrow of the United States Department of Agriculture and Walter Knott, a California berry expert, began looking into reports of a large, reddish-purple berry growing on farmland abandoned by farmer Rudolf Boysen. On Boysen's old farm, they found several berry vines growing among weeds, Darrow and Knott relocated the vines to Knott's farm and nursed them back to health. This new variety ultimately became known as boysenberries. Mrs. Knott began making preserves from these popular berries, which gave rise to world famous Knott's Berry Farm. Besides jams and jellies, boysenberries are also used to make syrups and pies.

Breadfruit

Select breadfruits that are lustreless, have an even dark green to brown color, are firm, uniform in shape, and heavy for their size. At full maturity, the skin will turn yellow, may have brown speckles, and a white sap will come to the surface and run over the outside. Unripe breadfruit may have a mildly offensive odor, which fades as it ripens.

Avoid breadfruits that are soft, cracked, bruised, or those that have black, moldy spots. Breadfruits that have a yellow stem may be overripe.

Store at room temperature up to two weeks. Ripe breadfruits should be eaten immediately or can be refrigerated for one day.

Peak Season: May – December

Nutritional Content: ½ cup breadfruit: 115 calories, 0 g fat, 28 g carbohydrates. Source of vitamin C, vitamin A, and potassium.

Did You Know...

- An excellent way of preparing breadfruit is to roast it over charcoal for about an hour, remove the heart, skin the fruit, and serve sliced.
- Breadfruit is an energy source and staple of South Pacific islanders. Its seeds, leaves, and blossoms are all edible.
- The fibrous inner bark of the breadfruit tree is used to make a traditional cloth called "Tapa," while the wood is used to make canoes and furniture, and its sap to make glue.
- Breadfruit trees grow up to sixty-feet tall with leaves ranging from one to three-feet long.
- The infamous Captain Bligh of the ship "Bounty" collected breadfruit cuttings in Tahiti for planting in the West Indies to feed slaves.

Cantaloupe

Select cantaloupes that are smooth, well formed, and golden color. They should be heavy for their size, and yield to palm pressure. Skin netting should be prominent, evenly distributed, coarse, and thick. The stem end must be smooth, slightly sunken, well rounded, and will emit a strong melon aroma when ripe. A bleached area where the melon rested on the ground is normal.

Avoid cantaloupes with large discolored areas or soft spots. Cantaloupes that are overly green were picked too early and will not ripen. A yellow color beneath the skin netting indicates the melon is overripe.

Store at room temperature for several days until ripe. Once ripe, refrigerate in a plastic bag up to one week.

Popular Varieties: American Pineapple and Boule d'Or.

Peak Season: May – September

Nutritional Content: ¼ medium cantaloupe: 50 calories, 0 g fat, 12 g carbohydrates, 11 g sugar. Good source of vitamin C, vitamin A, and potassium.

Did You Know...

- When shaken, an exceptionally juicy cantaloupe's seeds will rattle.
- If left unwrapped, a cantaloupe's aroma will permeate other foods in the refrigerator.
- Cantaloupes as we know them in America are actually muskmelons. True cantaloupes are mostly grown in Europe and have a hard, warty rind with deep grooves, as compared to the muskmelon's netted rind.
- Although cantaloupes got their name from the Italian town of Cantalupo near Rome, they originated in Iran and were cultivated by ancient Egyptians, Romans, and Greeks.

Carambola

Select carambolas that are light to dark yellow with no green tinges. They should be fairly firm with a smooth, shiny skin. Slight browning along the edges is normal. Although best if ripened on the tree, carambolas will ripen slowly if picked before full maturity. The fruit is easily damaged at all stages of ripeness; handle with care.

Avoid carambolas that are soft, blemished, or have excessive browning.

Store at room temperature until ripe, then refrigerate in a plastic bag up to one week. Slices may be frozen.

Popular Varieties: Arkin (four to five inches long, yellow to yellow-orange skin, sweet, juicy, firm flesh), Golden Star (large, deep winged, bright golden-yellow skin, very waxy, juicy, crisp, fiber-free flesh), Maha (Hawaii origin, roundish, light yellowish-white skin, crunchy, sweet, white flesh), Thayer and Newcombe (tart varieties, used in jams).

Peak Season: July – February

Nutritional Content: 1 medium carambola: 40 calories, 1 g fat, 8 g carbohydrates, 9 g sugar. High source of vitamin C, source of potassium.

Did You Know...

- Carambolas are considered a luxury item in supermarkets and are also known as Starfruit, Carambas, Five Fingers, and Five Corners.
- Carambola juice is used in making stain removers.
- Carambolas originated in Sri Lanka and the Maluku Islands, and have been cultivated in Southeast Asia and Malaysia for centuries. They were cherished by wealthy Europeans in the 1700s, and were introduced to Florida in 1887.
- Portuguese traders brought carambolas to South America and Africa. The word carambola means “food appetizer” in Portuguese.

Cherimoya

Select ripe cherimoyas that are brownish-green in color, heavy for their size, and yield to gentle palm pressure. Because they bruise easily, cherimoyas are often sold hard and green.

Avoid cherimoyas that are too soft or have brown, or bruised areas.

Store at room temperature away from direct sunlight for three to four days until skin softens and turns a speckled brownish color. Once ripe, wrap in a damp paper towel and refrigerate for one or two more days for optimal flavor.

Popular Varieties: Bays (round, medium size, light green skin, lemon-like flavor), Booth (medium size, seedy, papaya-like flavor), Honeyhart (medium size, smooth, yellowish-green skin, juicy, excellent flavor), McPherson (small to medium size, dark green skin, banana-like flavor), Whaley (medium to large size, light green skin, good flavor). There are twenty-six varieties.

Peak Season: November – May

Nutritional Content: ½ medium cherimoya: 257 calories, 1 g fat, 67 g carbohydrates. Good source of vitamin A and vitamin C, high source of fiber.

Did You Know...

- The ambrosia-like taste of cherimoyas resembles that of a pineapple, banana, papaya, coconut and mango all in one.
- Cherimoyas are also known as "custard apple" or "custard fruit," and were described by Mark Twain as "deliciousness itself."
- In order to increase the quantity of cherimoyas, they are pollinated by hand because bees have trouble flying into their flowers.
- Cherimoyas originated in the Andes Mountains, and were prized by Inca Emperors.
- Cherimoyas were first planted in 1871 in California, where today they rank third among all sub-tropical fruit crops.

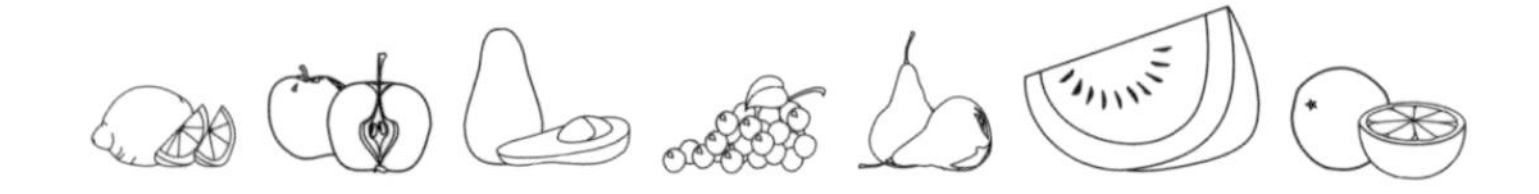

Cherry

Select cherries that are shiny, plump, and firm, but not overly hard. Depending on the variety, they should be brightly colored, ranging from yellow to deep mahogany-red. The deeper the color, the sweeter the cherries. Larger cherries are likely to be tastier; those with fresh green stems tend to last longer. Handle cherries carefully because they bruise easily.

Avoid cherries that are pale in color, bruised, moldy, shriveled, or those with dried-out stems. Skin cuts and missing stems promote decay. Check the bottom of the container; juice leakage indicates some cherries are smashed and possibly rotten.

Store unwashed cherries loosely packed in a plastic bag in the refrigerator up to one week for optimum freshness. Good air circulation helps preserve freshness. Discard soft, bruised cherries and refrigerate remainder as soon as possible. If properly stored, they will keep up to three weeks. Wash just before using and handle gently.

Popular Varieties: Two main groups: Sweet and Tart.

Sweet Varieties: Bing (red-mahogany color, juicy, crisp, firm with a small seed) and Lambert (dark red, smaller, heart shaped, rich, sweet flavor).

Tart Variety: Montmorency (bright scarlet, used mainly for making pies).

Peak Season: June – July

Nutritional Content: 1 cup cherries: 90 calories, 0.5 g fat, 22 g carbohydrates, 19 g sugar. Good source of vitamin C, vitamin A, and vitamin B.

Did You Know…

- You can substitute dried cherries for raisins in cookies, cakes, breads, and sauces.
- Studies show that cherry juice can help prevent tooth decay.
- When mixed with raw ground beef, cherry enzymes can help stop the development of harmful bacteria.
- Cherry pits discovered on lakeside cities in Switzerland date back to the Stone Age.
- In ancient Greece, physicians used cherries to treat epilepsy.
- Cherries originated in Asia Minor, got their name from the ancient Turkish city of "Cerasus," and were brought to America by early French settlers.
- Bing cherries were developed only one hundred years ago in the United States, and were named after a Chinese orchard worker.

Coconut

Select fresh coconuts that are heavy for their size, and when shaken sound full of liquid. The absence of liquid means the coconut is dried out.

Avoid coconuts that have mold, cracked shells, or damp eyes.

Store whole, unopened coconuts at room temperature up to six months. Once opened, refrigerate coconut meat in a tightly sealed container up to one week. Best when soaked in its own juice. Store canned coconut at room temperature up to eighteen months and bagged coconut up to six months. Once opened, refrigerate immediately.

Popular Varieties: Coconut Palm (most popular), Spiny Club Palm (small, sweet, red fruit, found in Costa Rica), Honey Palm (nut-like fruit, honey flavored sap, found in Chile).

Peak Season: Available year-round

Nutritional Content: 1 piece coconut, 2 x 2½ in.: 55 calories, 5 g fat, 2 g carbohydrates, 1 g sugar. Good source of potassium, source of fiber and iron.

Did You Know...

- You can use coconut milk to lightly flavor and texturize many soups, sauces, and desserts. Also use it to cook rice and pasta.
- By piercing one of a coconut's three eyes with an ice pick or screwdriver you can drink its inner liquid, which is an excellent beverage. This inner liquid is commonly mistaken for coconut milk; the milk actually comes from grating fresh coconut.
- Many beaches contain coconut trees because they always grow close to water. A coconut tree needs to absorb almost one hundred gallons of water to generate a single coconut.
- In ancient China, it was thought that coconuts sprang from decapitated heads.
- Coconuts got their name from early Portuguese explorers because of their resemblance to a face or "coco." The origin of coconuts is unknown.
- Coconuts were once so scarce in Europe that their shells were laced with gold and used to decorate palaces.

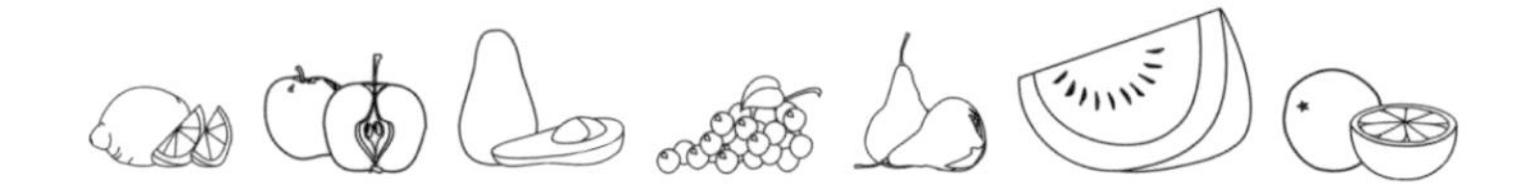

Cranberry

Select cranberries that are glossy, plump, and firm. They should be dry to the touch and intense in color ranging from light to dark red. Unlike other berries, cranberries are very durable and require no special handling. A good cranberry will bounce.

Avoid cranberries that are damp, discolored, shriveled, or soft.

Store in an airtight plastic bag in the refrigerator up to one month. They can be frozen up to one year. Do not store wet. Quickly rinse before using.

Popular Varieties: European, American, Mountain, Highbush.

Peak Season: September – January

Nutritional Content: 1 cup cranberries: 60 calories, O g fat, 14 g carbohydrates, 10 g sugar. Good source of vitamin C and fiber.

Did You Know...

- Cranberry juice consumption can help prevent bladder infections by increasing urine acidity.
- Most of the cranberries harvested each year are used for making sauce and jelly, or bottled for juice.
- Cranberries are one of only three fruits native to America.
- Cranberries were used by Native Americans as a meat sauce, to treat arrow wounds, and to dye blankets.
- Pilgrims used cranberries at the first Thanksgiving dinner with Native Americans and originally named them "craneberry" because their flower resembled the head of a sand crane.
- Many early voyaging ships stocked up on cranberries. Their high vitamin C content helped prevent scurvy, an ailment caused by a vitamin C deficiency.

Currant

Select berries that are firm, shiny, and bright in color.

Avoid currants that are soft, bruised, or moldy. Look for seeping juices. Seepage may be a sign that the berries are smashed or possibly rotten.

Store berries refrigerated up to three days. Red currants will usually keep longer than the black variety. Wash and gently remove stems before eating.

Popular Varieties: Red (tart, translucent, used for juice, jellies, and purees), Pink (colorless skin, pink flesh), White (nearly transparent skin, used for all cooking purposes), Black (brownish-purple skin, characteristic aroma, mainly used in pies, jams, and puddings).

Peak Season: July – August

Nutritional Content: 1 cup fresh currants: 84 calories, 0 g fat, 20 g carbohydrates, 11 g sugar. Good source of vitamin C.

Did You Know...

- Currants are good for jellies, pies, puddings, and sauces, and are also used to make wine and liqueur.
- Currant juice is used to make lozenges.
- Boiled currant juice can be used to treat inflamed sore throats.
- Currants have been cultivated since the 1500s in Europe and the 1700s in North America.
- These berries have been called currants since 1550 because of their resemblance to the dried currant raisins made from small seedless grapes from the Greek city of Corinth. Their older English name, "ribes," is still used in other languages.

Dates

Select dates that are plump, soft, and have a smooth, shiny skin. They should also be purchased in sealed packages. Dried dates will appear more wrinkled than fresh ones.

Avoid dates that are shriveled, moldy, or have sugar crystals on the skin. Also avoid those that have skin splits, a sour odor, or are rock hard.

Store fresh dates refrigerated and wrapped in plastic up to several weeks. Dried dates can be refrigerated up to one year.

Popular Varieties: Medjool (semisoft, large, smooth, rich flavor), Halawy (soft, thick flesh, sweet), Deglet Noor (semisoft, most popular, delicate flavor, firm texture), Thoory (driest variety, firm skin, chewy flesh). There are three categories of dates that vary according to softness: soft, semisoft, and dry. Note: Dry dates and dried dates are not the same thing: Dry dates are date varieties that contain less moisture when ripe; dried dates are dates that have been deliberately dehydrated after harvest.

Peak Season: July – October

Nutritional Content: 5-6 dates: 120 calories, 0 g fat, 21 g carbohydrates, 39 g sugar. Good source of fiber.

Did You Know...

- Dates, the oldest known fruit, are nutritious, fat-free, and packed with energy, making them a perfect candy snack.
- The Middle East is the principal date-growing region in the world and its largest consumer.
- The date tree symbolizes victory and justice because its wood does not decay.
- Because of their shape, it is believed dates got their name from the Greek word meaning finger.
- The Medjool date, which was considered a delicacy and the "Fruit of Kings," was imported to the United States from French Morocco in the 1920s and is now exported from the United States to several Middle Eastern countries.

Fig

Select figs that are richly colored, soft, but not mushy, and have a slightly fragrant odor. Figs are fully ripened and partially dried on the tree. Depending on the variety, the color may range from purplish-brown to greenish-yellow. They should be purchased in sealed packages.

Avoid figs that are bruised, moldy, or have broken skins. A sour odor is a sign of juice fermentation, indicating spoiled fruit.

Store fresh figs in the refrigerator up to two or three days. Dried figs can be stored either in the refrigerator or at room temperature and will keep for several months.

Popular Varieties: Black Jack (large, purplish-brown skin, sweet, juicy, strawberry-red flesh), Celestial (purplish-brown skin, pink flesh), Janice Dakota (large, yellow skin, almost seedless), White Genoa (greenish-yellow skin, amber flesh, distinctive flavor).

Peak Season: Available year-round

Nutritional Content: 2 dried figs: 100 calories, 0 g fat, 23 g carbohydrates, 25 g sugar. Great source of fiber and has the highest overall mineral content of all common fruits.

Did You Know...

- Figs contain more potassium than a banana and more calcium than skim milk.
- Fig blossoms don't grow on branches, but inside the fruit. These flowers create the crispy small seeds that give figs their characteristic texture.
- Eating figs is part of traditional American and Jewish holidays such as Passover, Succoth, and Hanukkah.
- Figs originated in the Mediterranean region, and were so prized by ancient Greeks that it was illegal to export them.
- Fig wreaths were the first medals awarded at the Olympics.
- The fig is the Bible's most-mentioned fruit and was possibly the first to be dried and stored by man. Figs have been found at excavation sites dating back to 5,000 B.C., and were a favorite of Cleopatra.

Grape

Select grape clusters that are plump, fairly firm, and fully colored. They should be securely attached to green stems. When ripe, dark grapes should have a deep purple-black color with no sign of green; green grapes, a slight yellow or straw hue; and red varieties, a predominantly rich red color. A powdery bloom is an important sign of freshness. Be selective because grapes do not ripen once picked. Tasting a grape is the best way to confirm its sweetness.

Avoid grapes that are soft, wrinkled, or bruised, indicating they may be old or have not been properly refrigerated. Also avoid those with shriveled brown stems or stems encircled by a bleached area.

Store unwashed grapes refrigerated in a plastic bag up to one week. They are quite perishable. Thoroughly wash them before eating.

Popular Varieties: Flame Seedless (red variety, crunchy texture, sweet, tart flavor), Red Globe (red variety, crisp, delicate flavor), Green Seedless Thompson (green variety, crisp, sweet flavor), Concord (black variety, sweet, tart flavor).

Peak Season: June – November

Nutritional Content: 1½ cups grapes: 90 calories, 1 g fat, 24 g carbohydrates, 23 g sugar. Good source of vitamin C, source of vitamin A.

Did You Know...

- Letting grapes sit at room temperature for a few minutes will enhance their flavor.
- A substance found in grape skin, "Resvesterol," may be a strong cancer fighter and its flavonoids are known to deter plaque and possibly prevent the occurrence of heart attacks.
- The frosty look on grape varieties is called bloom, which provides a natural protection for the fruit.
- Grapes were a staple of early civilizations, decorated the tables of the ancient Pharaohs, and were used as a form of currency by Mediterranean traders.
- In the late 1700s, Franciscan missionaries journeyed north from Mexico and cultivated the first vineyards in California as a source of sacramental wine.

Grapefruit

Select grapefruits that are glossy, brightly colored, and evenly shaped with slightly flattened ends. They should be mildly fragrant, firm, yet yield to palm pressure. Those that are heavy for their size with a thin, fine-textured skin are the juiciest. Rolling them between your palm and a flat surface for a few seconds just before eating will make them juicier.

Avoid grapefruits that are pointed on the stem end, indicating a thick skin; these tend to be less sweet. Coarse, thick-skinned grapefruits will have less juice.

Store ripe grapefruits refrigerated in a plastic bag up to six weeks. They are quite perishable, so refrigerate immediately. Fruit will be juicier if left at room temperature before eating or juicing.

Popular Varieties: White (seedless) and Pink (higher source of vitamin A).

Peak Season: January – June

Nutritional Content: ½ medium grapefruit: 60 calories, 0 g fat, 16 g carbohydrates, 10 g sugar. Excellent source of vitamin C and fiber, source of vitamin A.

Did You Know...

- Peeling grapefruits is easier if they are first placed in boiling water.
- Grapefruits can be sweetened by sprinkling a little salt on the flesh.
- Grapefruits got their name from the way they grow, in clusters, like grapes.
- Grapefruits were initially used by early Floridians to create a tonic for treating general health ailments, including influenza.
- Grapefruits are native to the tropics of Barbados, and were the result of a natural cross-pollination between the orange and pomelo.

Guava

Select guavas that are firm, fragrant, and vary in color from white to dark red. A green, yellow-white or deep purple thin skin indicates ripeness. Brown veins, scars, scuff marks, or spots are natural and will not affect the quality. If you are not planning to use them immediately, select guavas at a mature green stage.

Avoid guavas that are overly soft, have discolored soft spots and/or split or broken skins.

Store at room temperature until ready to eat. They will ripen faster if stored in a paper bag. Once ripe, they will yield to gentle pressure and can be refrigerated up to one week.

Popular Varieties: White (commonly cultivated), Yellow (commonly cultivated), Strawberry (wine colored, flavor resembles that of a strawberry). There are more than one hundred fifty varieties.

Peak Season: June – March

Nutritional Content: 1 medium guava: 30 calories, 0.5 g fat, 6 g carbohydrates, 5 g sugar. Rich source of vitamin C, good source of fiber.

Did You Know…

- When making guava puree, be sure to include the rind, which when combined with the pulp has up to five times more vitamin C than an orange.
- Guavas, classified as berries, are one of the few tropical fruits that contain pectin, which is essential for making jams, jellies, and marmalades.
- During the harvest period, guava trees must be picked an average of thirty-five times because guavas ripen at different rates.
- Guavas derive their name from the Spanish word "guayabe" meaning guava tree.

Honeydew

Select honeydew melons that have a waxy, velvety-smooth skin. Depending on the variety, color can range from creamy white to pale yellow. They should be well shaped and heavy for their size. The stem end of the fruit should be fragrant and yield to gentle pressure when ripe. Those weighing roughly five pounds will have the best flavor.

Avoid melons that are greenish in color or too hard, indicating they were possibly picked too early and may not fully ripen. Also avoid those that are fuzzy, have surface defects, broken rinds, or soft spots.

Store melons at room temperature until the blossom end softens. Wrap tightly and refrigerate up to one week.

Popular Varieties: Honeydew (creamy white or yellow, large, strong aroma) and Golden Honeydew (vibrant gold color, thin rind, very sweet).

Peak Season: June – October

Nutritional Content: 1⁄10 medium honeydew: 50 calories, 0 g fat, 13 g carbohydrates, 12 g sugar. Excellent source of vitamin C.

Did You Know…

- Honeydews are the sweetest of all melons.
- Honeydew melons are second only to bananas as the most consumed fresh fruit in the United States.
- If the seeds rattle when you shake a honeydew, the fruit is exceptionally juicy.
- Due to their high sugar content and fragile nature, honeydews can only be harvested before they are fully ripe.
- Honeydews can be traced back to Asia, and were cherished by ancient Egyptians.

Kiwi Fruit

Select kiwis that are large, plump, and egg shaped. They should be firm, blemish free, fragrant, and yield to slight finger pressure, indicating ripeness. Kiwis are very delicate and must be handled carefully to avoid bruising.

Avoid kiwis that are wrinkled, bruised, too soft, or have skin splits.

Store at room temperature until they yield to slight finger pressure, or refrigerate up to two weeks. Kiwis are highly perishable and rot easily.

Popular Varieties: New Zealand Hayward (most common), Alison, Bruno, Gracie, Monty, Matua, Chico.

Peak Season: Available year-round

Nutritional Content: 1 medium kiwi: 50 calories, 0.5 g fat, 12 g carbohydrates, 8 g sugar. Excellent source of vitamin C, good source of vitamin E, fiber, and potassium, source of vitamin A.

Did You Know…

- There are four hundred varieties of kiwi fruit in China.
- Although kiwi fruit is more than seven hundred years old, it has been available in the United States only since 1962.
- It is believed kiwi fruit was named for New Zealand's national bird, the kiwi, which has a fuzzy brown exterior resembling the kiwi skin.
- Kiwi fruit is not native to New Zealand, but originated in the Chang Kiang Valley of China, where it was regarded as a delicacy by the court of the great Khans. For many years it was known as the "Chinese Gooseberry." Chinese missionaries introduced kiwis to New Zealand at the turn of the twentieth century.

Kumquat

Select kumquats that are plump, firm, and heavy for their size. They should have a shiny, uniform, deep golden-orange color. The entire fruit—including its seeds, juicy, sweet skin, and its dry, somewhat tart flesh—is edible.

Avoid kumquats that are green tinged; this may indicate that they are tart and not yet ripe. Also avoid kumquats that are blemished, overly soft, wrinkled, or too small.

Store refrigerated up to several weeks. Freezing is not recommended.

Peak Season: December – May

Nutritional Content: 8 kumquats: 100 calories, 0 g fat, 25 g carbohydrates. High source of vitamin C and fiber, source of vitamin A and potassium.

Did You Know...

- Although kumquats can be eaten raw, the taste will be slightly tart. They are most commonly used in making syrups and preserves. They also taste great candied.
- Their name comes from Cantonese meaning "golden orange" and originally was spelled kam kwat.
- Kumquats, a citrus fruit, are native to China. Although grown for many years in both China and Japan, they were introduced to Europe only one hundred fifty years ago after being discovered growing in China in 1848 by Robert Fortune.

Lemon

Select lemons that are plump, firm, and heavy for their size. The skin should be smooth, thin, slightly shiny, and have an even, yellow color. Medium to large lemons are generally the better choice.

Avoid lemons that are hard, spongy, and soft skinned, or that have a dark yellow skin, indicating they are old. Coarse, thick-skinned lemons will have less juice.

Store at room temperature in a cool place up to one week. They can also be refrigerated in a plastic bag up to three weeks. Lemons stored at room temperature will yield more juice. Slice them right before using to avoid exposing them to oxygen, which lessens their vitamin C content.

Popular Varieties: Two basic types: Acid (supermarket variety) and Sweet (home garden and ornamental variety). Varieties: Lisbon (acidic type), Eureka (acidic type), Meyer (sweet type).

Peak Season: Available year-round

Nutritional Content: 1 medium lemon: 15 calories, 0 g fat, 5 g carbohydrates, 1 g sugar. Good source of vitamin C.

Did You Know...

- Lemons can be made juicier by rolling them between your palm and a flat surface.
- Microwaving a lemon for thirty seconds will intensify its juice content.
- Lemon juice will help prevent sliced apples, avocados, and pears from browning, but will also give them a bit of a tart taste.
- Odors can be removed from pots, pans, and hands by rubbing them with a cut lemon just before washing.
- Lemon peels can keep a garbage disposal smelling fresh.
- Lemonade is not only a great thirst quencher, but also helps in reducing fever. Mongolians invented lemonade in 1299.
- California Gold Rush miners paid a dollar each for lemons in order to prevent scurvy, an ailment caused by a vitamin C deficiency. At this price, a glass of lemonade would cost $37 in today's economy.

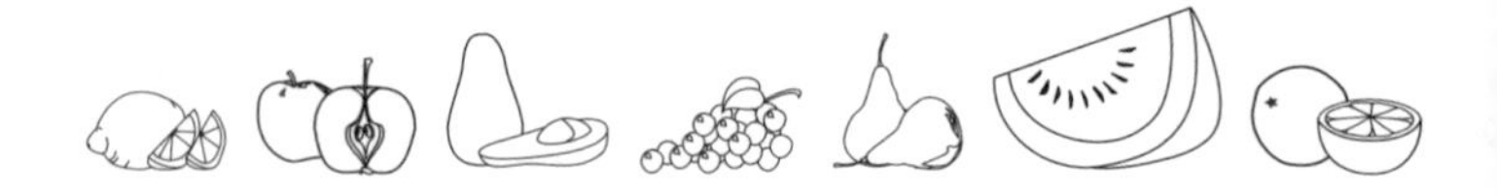

Lime

Select limes that are glossy and light to deep green in color. They should have a thin, smooth skin and be heavy for their size. Small brown areas on the skin will not affect the flavor. They must be firm, but not too hard.

Avoid limes that have a yellowish skin or are too small. A hard, shriveled skin is a sign of dryness. Coarse, thick-skinned limes will have less juice.

Store at room temperature or refrigerated in a plastic bag up to three weeks. They must also be kept out of direct sunlight, which will cause them to shrivel and become discolored. Limes stored at room temperature will yield more juice.

Popular Varieties: Tahiti or Persian (most popular) and Key Limes (smaller and rounder, more acidic, used in pies).

Peak Season: Available year-round

Nutritional Content: 1 medium lime: 20 calories, 0 g fat, 7 g carbohydrates, 0 g sugar. Excellent source of vitamin C.

Did You Know...

- Lime juice will remove fruit juice stains from countertops and fabrics.
- The somewhat acidic flavor of lime juice is a great salt substitute for those on a low-sodium diet.
- Unlike lemon juice, lime juice does not pass on its taste to other foods, but rather complements their natural flavors.
- British sailors were called "limeys" because of their high lime diet, which was rich in vitamin C and aided in preventing scurvy, an ailment caused by a vitamin C deficiency.
- Christopher Columbus is supposedly responsible for starting the first lime orchards on the island of Hispaniola. Spanish conquistadors later planted them in Florida.

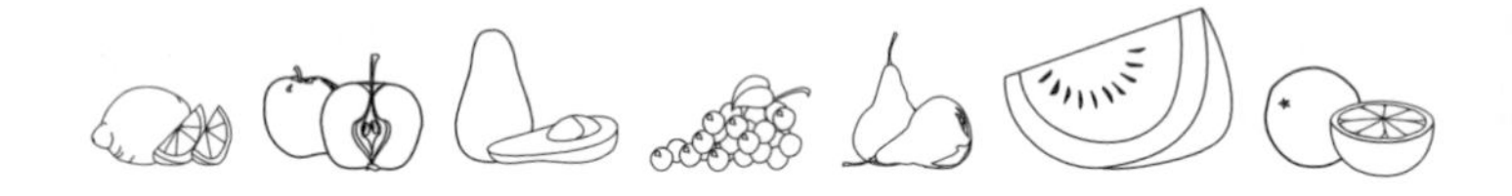

Loganberry

Select loganberries with a glossy, deep, bright red color with no hulls attached. Berries should be plump, uniform in size, and free of seeping juices.

Avoid loganberries that are bruised, shriveled, moldy, or have green caps still attached, indicating they were picked too early. Wet berries are a sign of damage or possible decay.

Store unwashed berries refrigerated in a single layer up to two or three days. Wash prior to using and handle gently; berries bruise easily.

Peak Season: June – August

Popular Varieties: Hybrid

Nutritional Content: 1 cup loganberries: 89 calories, 1 g fat, 22 g carbohydrates, 11 g sugar. Good source of vitamin A and potassium.

Did You Know...

- Loganberries, a hybrid between a red raspberry and a wild blackberry, are mainly grown for making wine, juice, and pies. Their sweet, tart flavor is often used to complement sweeter berries.
- Willamette Valley in Oregon accounts for more than ninety-five percent of the domestic production.
- Judge James Harvey Logan, a backyard farmer, is recognized for creating the loganberry in about 1916 in an attempt to develop better commercial berry varieties. Unlike most other berries, they do not exist in the wild.

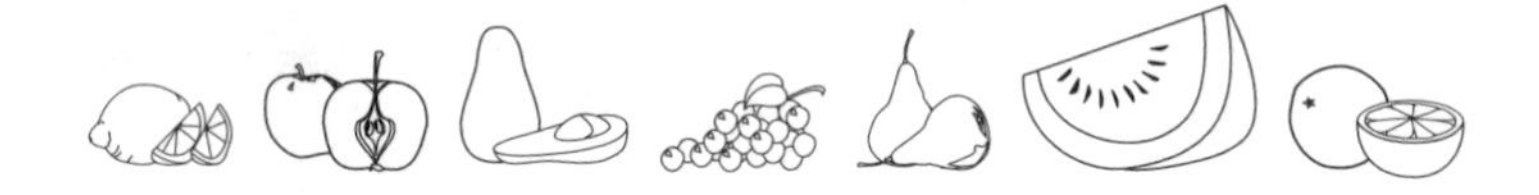

Lychee

Select lychees that are firm, plump, fragrant, and heavy for their size. Those that are plump near the stem area sometimes have a smaller seed, which means more fruit. The redder the shell, the fresher the lychee.

Avoid lychees that are too brown or black, indicating dryness, and those that are split or soft.

Store refrigerated up to two weeks..

Popular Varieties: Amboina (medium size, bright red), Hak ip (medium size, red with green tinge, soft skin, crisp flesh), Kwa luk (large, red with green tips, fragrant), Kwai mi (medium size, slightly oval, reddish-brown, firm), Tai tsao (bright red, egg shaped, rough skin, crisp, sweet, firm flesh). There are thirty-three varieties.

Peak Season: June – July (found fresh mostly in Asian markets)

Nutritional Content: 15 lychees: 100 calories, 1 g fat, 24 g carbohydrates, 0 g sugar. Excellent source of vitamin C, potassium, and phosphorous.

Did You Know...

- Lychees have been used to relieve coughs and enlarged glands.
- The United States Department of Agriculture and the National Cancer Institute are reportedly testing lychee roots for tumor fighting properties.
- Even though lychees are also called lychee nuts, they are not nuts and their seeds are not edible.
- Although lychees hail from southern China and are glorified and portrayed in Chinese literature and art, they are also grown on a small scale in Hawaii, California, and Florida.
- Lychee trees can live for more than one hundred years.
- The first book ever written on fruit in 1056 was about lychees.

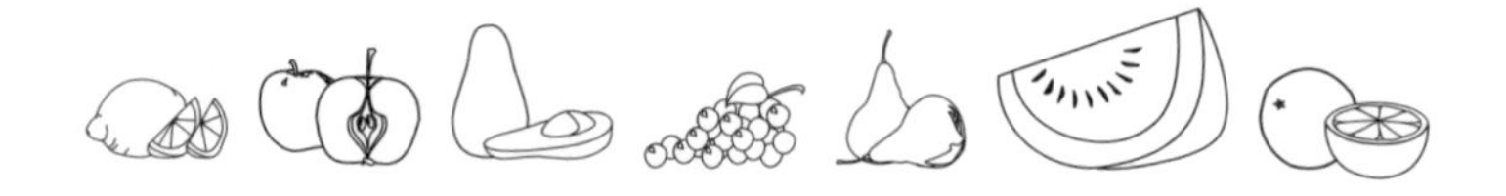

Mandarin Orange

Select Mandarin oranges that have a glossy, bright orange color, a bit lighter than a tangerine. They must be smooth skinned and heavy for their size. Skin should be tighter than that of a tangerine and fruit will range from firm to slightly soft.

Avoid Mandarin oranges that have noticeable blemishes, soft spots, are moldy, or have a dull, rough, or bumpy skin.

Store at room temperature for a week or so away from direct sunlight or heat, or refrigerate in a plastic bag up to two weeks.

Popular Varieties: Kinnow, Royal, Satsuma, Sunburst.

Peak Season: January – May

Nutritional Content: 1 medium Mandarin orange: 45 calories, 1 g fat, 16 g carbohydrates, 8 g sugar. High source of vitamin C, good source of fiber.

Did You Know...

- Mandarin oranges are a cross between a tangerine and an orange.
- Mandarin oranges are native to Southeast Asia.
- Australia is the largest consumer of Mandarin oranges.
- Mandarins got their name from the elite upper class of the Chinese Empire for whom they were exclusively reserved.
- The oil extracted from Mandarin peels is used commercially in flavoring candy, ice cream, chewing gum, liqueurs, and carbonated beverages. Countries such as Italy, Sicily, and Algiers commonly use Mandarin peel oil in making perfume.

Mango

Select mangoes that are unblemished, plump, and firm. They should be fragrant, and yield to slight palm pressure when ripe. Depending on the variety, their color can range from completely green to canary yellow blushed with red. Mango size and color differ according to variety, but is not an indicator of quality or ripeness. Larger mangoes have a higher fruit-to-seed ratio.

Avoid mangoes that have a grayish skin discoloration, soft bruised spots, or are shriveled. Black speckling is normal as mangoes ripen.

Store unripe mangoes at room temperature. If stored below fifty degrees, much of their flavor will be lost. Once ripe, they can be refrigerated up to one week. Sliced mangoes should be wrapped in plastic and refrigerated.

Popular Varieties: Tommy Adkins (medium-large, oval shaped, bright red to orange skin), Keitts (largest variety, greener with a yellow or rose blush, less fibrous), Kent (large, irregular oval shaped, orange-yellow with a slightly red blush), Haden (oval shaped, yellow in color, fibrous).

Peak Season: May – August, November – December

Nutritional Content: 1 medium mango: 34 calories, 1 g fat, 34 g carbohydrates, 30 g sugar. Rich source of vitamin C, vitamin A, and beta-carotene.

Did You Know...

- You must be careful when eating a mango; the juice will stain clothing.
- Mangoes are the world's most popular fruit and although they are the most favorite tropical fruit in the United States, they are also its most under-rated fresh fruit.
- Over the years, the mango tree became a sign of good luck and wealth.
- Asian legend maintains that a wish made under the shade of a mango tree will come true.
- Mangoes originated in Burma and Asia, and have been cultivated for more than six thousand years. In ancient India, they were known as the "Food of the Gods."
- Mango twigs were used as early toothbrushes. Chew sticks are made from their leaves and twigs in India and Panama.

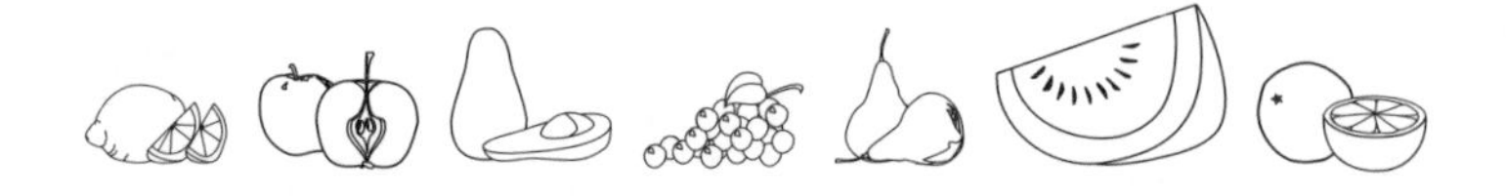

Nectarine

Select nectarines that have a bright golden-yellow color and are fragrant. They should be firm, but not rock hard, yielding to gentle palm pressure on their seams. Handle them carefully because many varieties are fragile. Nectarines are already ripe when picked and do not ripen well after they are harvested.

Avoid nectarines with skins that are punctured, shriveled, cracked, or those that are too small or hard. Flat, brownish bruises are signs of decay.

Store under-ripe nectarines at room temperature for a couple of days. Keep them out of the refrigerator until they are ripe to avoid a mealy consistency. Ripe nectarines will stay fresh in the refrigerator an average of five days.

Popular Varieties: Two basic types: Clingstone (fruit clings to the pit) and Freestone (fruit removes easily from the pit). Varieties: Arctic Glo (sweet, white flesh), Arctic Queen (rich flavor, crunchy texture, white flesh), Arctic Star (dark red skin, semi-freestone flesh).

Peak Season: June – August

Nutritional Content: 1 medium nectarine: 70 calories, 0.5 g fat, 16 g carbohydrates, 12 g sugar. High source of vitamin C, source of vitamin A.

Did You Know...

- The only major difference between the appearance of a nectarine and a peach is the nectarine's smooth skin versus the peach's fuzzy surface. Nectarines are also sweeter than peaches and keep for a longer period of time because of their firm flesh.
- Nectarines are a type of peach. Contrary to popular belief, they are not a cross between a peach and a plum.
- Nectarines received their name from the Hellenic description of their flavor as being like the "nectar of the gods." The Greek word "nektar" means "drink of the gods."
- Nectarines originated in China, and reached Greece and Rome via trade routes.

Orange

Select oranges that are firm, heavy for their size, and have a smoothly textured, glossy skin. A rough, brownish area on the skin or slight greening should not affect the quality. It is best to purchase loose rather than pre-bagged oranges to avoid getting bad or under-ripe fruit. Small to medium varieties tend to be sweeter than larger varieties; thin-skinned varieties tend to be sweeter than thick-skinned varieties.

Avoid oranges that have indentations and soft or moldy spots. Because they are often dyed, a bright, even color does not necessarily mean quality or ripeness.

Store at a cool room temperature for several days. Oranges are hardy and will last in the refrigerator for up to two weeks.

Popular Varieties: Valencia (best for juicing) and Navel (two fruits in one, the characteristic "belly button" is actually another tiny orange growing at the tip of the larger one).

Peak Season: Available year-round

Nutritional Content: 1 medium orange: 70 calories, 0 g fat, 21 g carbohydrates, 14 g sugar. Excellent source of vitamin C and fiber, source of vitamin A and potassium.

Did You Know...

- Frozen concentrate, fresh squeezed, and pasteurized orange juice all contain the same amount of vitamin C.
- Although an orange tree takes up to five years to produce fruit, once it begins to flourish it can generate more than one thousand pounds of oranges a year for as long as fifty years.
- The residents of Hong Kong hold the record for orange consumption, using an average of fifty pounds of oranges per person per year.
- Oranges are believed to have originated in the fertile lands between the Tigris and Euphrates Rivers in what is now Iraq.
- Oranges were cultivated before the Middle Ages.
- Ancient Egyptians used orange juice as an embalming fluid.
- A fur trapper named William Wolfskill is renowned as being the first to grow Valencia oranges in California.

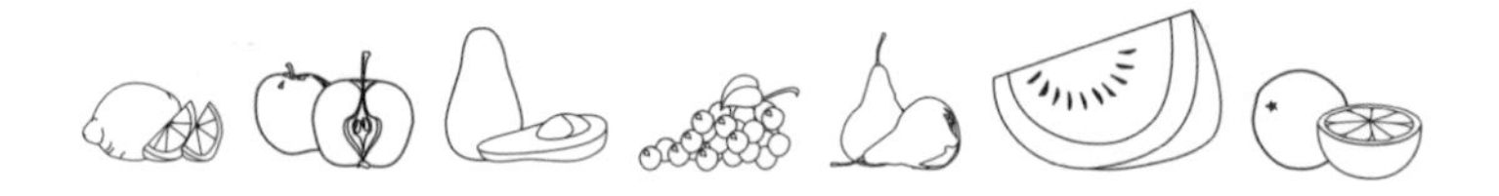

Papaya

Select papayas that are fairly large and have a smooth, two-toned skin with shades of yellow and green. They should yield to gentle pressure when ripe, and the skin should be at least half yellow. Completely green papayas were picked too early and may not ripen properly. Handle carefully to avoid bruising.

Avoid papayas that are blemished, broken, bruised, or too soft. The latter is an indication that papayas are overripe.

Store at room temperature until ripe, then refrigerate for a couple of days. Never refrigerate a papaya that is less than half ripe. Cold temperatures stop the ripening process. Papayas are extremely perishable.

Popular Varieties: Solo (most common, about six-inches long, vibrant yellow-orange flesh) and Maradol (weigh up to twenty pounds, salmon colored). There are dozens of varieties.

Peak Season: July – December

Nutritional Content: ½ medium papaya: 70 calories, 0 g fat, 19 g carbohydrates, 9 g sugar. High source of vitamin C and fiber, source of vitamin A and potassium.

Did You Know...

- You can tenderize meat by rubbing it with pureed papaya, or by soaking it in papaya juice.
- Papayas are known as the perfect fruit and have been used to alleviate indigestion, toothaches, fevers, and to remove freckles.
- The Solo papaya was so named because it was a perfect one-person snack.
- Botanists consider the papaya a berry.
- Hawaii accounts for the majority of the domestic production generating more than fourteen thousand tons of papayas each year with approximately half exported to Japan and the remainder shipped to the U.S. mainland.
- Papayas are indigenous to Central and South America, where pre-Columbian Indians first farmed them.

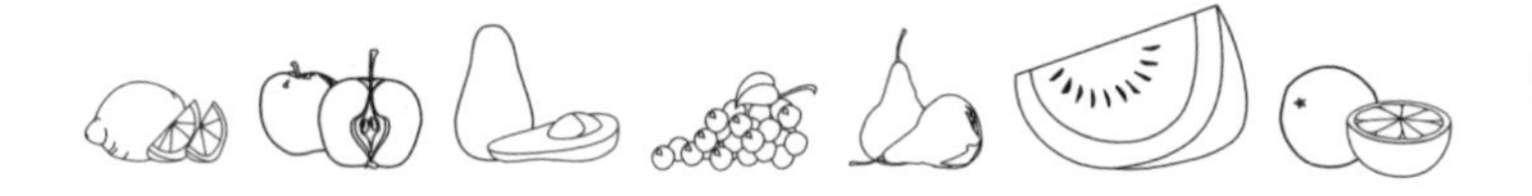

Passion Fruit

Select passion fruits that are large, firm, and heavy for their size. For the purple variety, the skin should be brown and wrinkled, indicating ripeness. The skin of the yellow variety will be an even, deep yellow color. Mold spots do not affect their quality and can be cleaned off.

Avoid passion fruits that are soft.

Store passion fruits at room temperature up to two weeks or until ripe, then refrigerate up to one week.

Popular Varieties: Two types: Purple (Black Knight, Edgehill, Frederick, Kahuna, Paul Ecke, Purple Giant, Red Rover) and Yellow (Brazilian Golden, Golden Giant). Purple varieties are sold fresh in markets. Yellow varieties are used for juicing. There are fifty edible varieties.

Peak Season: Available year-round

Nutritional Content: 2 passion fruits: 35 calories, 0 g fat, 8 g carbohydrates, 5 g sugar. Good source of vitamin A, vitamin C, and fiber, source of potassium.

Did You Know...

- The leaves of passion fruit vines are used by Amazon tribes and Native Americans as a sedative, and by Brazilians to treat asthma, bronchitis, and whooping cough. Peruvians use its juice to treat bladder infections.
- Hawaii has the highest consumption of passion fruit juice per capita in the world. Australia has the highest overall consumption.
- Passion fruit vines adorn many Hawaiian home gardens.
- Passion fruit flowers are called the "perfect flower." They open in the morning and close at night.
- Passion fruits are native to the Amazons, and were so named by Spanish missionaries because their blossoms resembled the nails and thorns of the Crucifixion.

Peach

Select peaches that are plump, and have a healthy cream to golden-yellow undertone. Ripe peaches will be fragrant and yield to slight palm pressure. A fully developed peach should have a crease running from stem to point. A reddish blush may be present on some varieties, but is not a true sign of quality or ripeness. Peaches do not get sweeter once picked, but will get softer and juicier. Handle them carefully to avoid bruising.

Avoid peaches that show signs of greening because they will not ripen, but simply soften and gain nothing in flavor. Also avoid those that are too hard or have soft spots, which indicate bruising.

Store at room temperature until ripe. To speed the ripening process, store in a paper bag. Peaches emit a gas called ethylene, a natural ripening hormone; the paper bag keeps the gases close to the fruit. Check daily because peaches ripen very quickly. Once ripe, refrigerate uncovered for a few days. Do not store them in a plastic bag; this tends to build moisture and promote decay. Their flavor is enhanced at room temperature, so take them out of the refrigerator at least thirty minutes prior to eating.

Popular Varieties: Two types: Freestone (fruit slips away easily from the pit, making slicing simple) and Clingstone (fruit adheres to the pit).

Peak Season: May – August

Nutritional Content: 1 medium peach: 40 calories, 0 g fat, 10 g carbohydrates, 9 g sugar. Source of vitamin A, vitamin C, and potassium.

Did You Know...

- Peaches should always be peeled before using them to make pies or preserves. Once cooked, their skins will become tough, and ruin the texture. To peel peaches, dip them in boiling water for twenty to thirty seconds and then immediately plunge them into a bowl of chilled water.
- More than three hundred varieties of peaches are grown in the United States, making them the country's third most popular fruit. The U.S. peach crop is larger than that of all other countries of the world combined.
- Peach pits occasionally produce nectarine trees and vice versa.
- Peaches are native to China, and were prized as a symbol of long life. In ancient times, one would have to journey to China or Persia in order to taste a peach, then known as "Persian Apples." The Chinese developed more than four hundred varieties of peaches, some shaped like a pancake.
- Offering a peach to someone in ancient Rome was a sign of friendship, possibly the origin of the expression "you're a real peach."

Pear

Select pears that are evenly colored and slightly fragrant. Minor surface blemishes or scars usually do not affect the inner quality. Pears are at their ripest, most flavorful stage when they yield slightly to gentle pressure near the base of the stem.

Avoid pears that are shriveled or have cuts, bruises, or soft spots, which are signs of interior discoloration. Handle them gently. Pears bruise easily.

Store at room temperature and check daily. Green, firm pears take four to six days to ripen. Once they yield to gentle pressure, refrigerate for a few days. Never store pears in plastic; they need air circulation.

Popular Varieties: Bartlett (most popular, bell shaped, summer variety, in season July – December), Anjou (oval shaped, winter variety, in season October – June), Bosc (long necked, in season August – May). There are more than three thousand varieties.

Peak Season: Different varieties are available year-round.

Nutritional Content: 1 medium pear: 100 calories, 1 g fat, 25 g carbohydrates, 17 g sugar. Source of vitamin C, fiber, and potassium.

Did You Know...

- Pears should always be peeled before using them in cooked dishes because the skin darkens and toughens when heated.
- Pears are thought to contain enzymes that promote relaxation by temporarily alleviating stress.
- Pears are members of the rose family and cousins to apples.
- The first pear tree was planted in America in 1620.
- Eating raw pears was once considered dangerous after noted historians and scholars claimed they were poisonous. Only in the eighth century, after Emperor Charlemagne of France grew them in his kingdom, did the delusion fade.
- Although pears are known as "the aristocrat" of fruit and have been consumed since the Stone Age, they were never mentioned in the Bible.

Persimmon

Select persimmons that are plump, and have a smooth, glossy skin, with green caps firmly attached. Their color can range from light yellow-orange to dark orange-red. When ripe, the Fuyu variety remains firm and crisp. The Hachiya variety will be soft, but not mushy when ready to eat.

Avoid persimmons with soft skins, skin splits, dark scars, or bruises.

Store refrigerated in a plastic bag up to three days. Cold temperatures make them sweeter. They will ripen quicker if placed in a paper bag with an apple or banana for a few days. Overripe persimmons will turn mushy quickly; watch them carefully.

Popular Varieties: Fuyu (non-astringent variety, flatter, tomato shaped) and Hachiya (astringent variety, large, round with a pointed base).

Peak Season: October – November

Nutritional Content: 1 large persimmon: 110 calories, 0 g fat, 31 g carbohydrates, 25 g sugar. High source of vitamin A, source of vitamin C.

Did You Know...

- You can make persimmons sweeter by keeping them in cold temperatures.
- Depending on the variety, persimmons are native to Japan, China, Mexico, and the United States.
- The wood of a persimmon tree is one of the hardest woods known, and is used in making golf clubs.
- According to folklore, the kernel of a persimmon seed can be used to forecast winter weather: spoon shaped means heavy snow; fork shaped, a mild winter; and knife shaped, a cold winter.
- Persimmons, also known as kaki or Sharon fruit, were introduced to California in the 1800s.
- Persimmons have been cultivated for centuries. There are more than two thousand varieties.

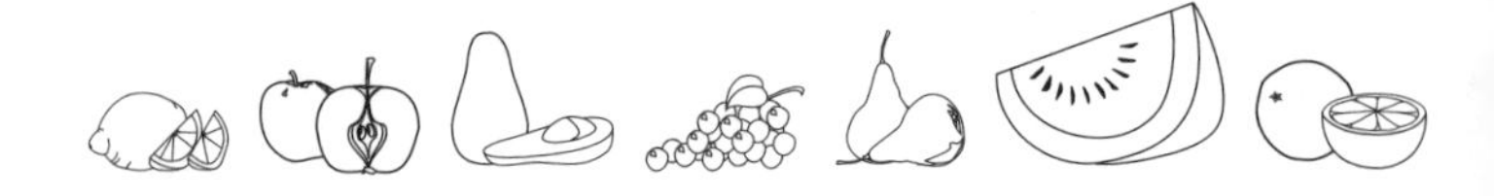

Pineapple

Select pineapples that are firm, plump, and have a full, vibrant yellow color. The more yellow the pineapple, the higher its sugar content. Pineapples with larger eyes (eye-shaped markings on the skin) tend to be sweeter. They should be slightly soft to the touch and fragrant on the underside. The crown of leaves must be crisp and green with no yellow or brown tips. Pineapples are already ripe when picked. Those with "jet-shipped from Hawaii" labels are best.

Avoid pineapples with dark, soft, or sunken skin areas. If they are not picked ripe, the starch will not convert to sugar and the fruit will never really ripen, so avoid signs of greening. Liquid leaking from the bottom of a pineapple or an unpleasant odor is a sign of fruit decay.

Store whole, ripe pineapples refrigerated and tightly wrapped up to three days, never at room temperature. Cut pineapple that is tightly sealed will last a few days longer.

Popular Varieties: Smooth Cayenne (most popular, best tasting, cone-shaped Hawaiian pineapple, sold fresh and canned), Red Spanish (squarish shape, tougher shell, most are sold fresh, grown in the Caribbean),

Sugar Loaf (large, most are sold fresh, imported from Mexico).

Peak Season: March – June

Nutritional Content: 2 slices pineapple: 60 calories, 0 g fat, 16 g carbohydrates, 13 g sugar. Good source of vitamin C.

Did You Know...

- Pineapples have their own powerful tenderizer, an enzyme called bromelain. Its juice can either be added to a marinade sauce, or stand alone as a marinade. Pineapple flavor best complements pork.
- It takes eighteen to twenty-two months to produce a single pineapple.
- The fruit gets its name from the Spanish word "piña" meaning cone. Pineapples are still called "piñas" in Latin America.
- Pineapples are a symbol of hospitality, and were once so rare they were called "the fruit of the kings."
- Historians believe voyagers brought pineapples to Hawaii from Tahiti and other South Sea islands around 1800.

Plantain

Select plantains that are green to almost black, depending on the desired taste and texture. Ripe plantains should be fairly firm and resemble elongated overripe bananas with spots and scars.

Avoid plantains that are soft, completely black, or have moldy spots.

Store uncovered at room temperature away from direct sunlight until ripe. They can be refrigerated until skin is completely black and slightly wilted.

Popular Varieties: Horn and French. Plantain is a starchy fruit and cousin of the banana.

Peak Season: Available year-round.

Nutritional Content: 1 cup raw slices: 180 calories, 1 g fat, 47 g carbohydrates, 8 g sugar. Good source of vitamin C, vitamin A, fiber, and potassium.

Did You Know...

- Although plantains are fruit, they are rarely eaten raw, but rather fried, baked, or boiled, similar to potatoes.
- Plantains are starchier than bananas because they contain eighteen percent less moisture. The higher the moisture content, the faster starch turns into sugar.
- South American Indians boiled plantain leaves and drank the liquid as a cure for colds. They still use plantain soup to treat colds and tuberculosis.
- In parts of East Africa, plantain is not only a staple food but is also used to make an intoxicating drink.
- Plantain leaves are used to make mats, bags, and cigarette paper.

Plum

Select plums that are plump, smooth, and well colored for their variety. Ripe plums will yield to gentle pressure and be slightly soft on the stem and tip. Depending on the variety, color ranges from red to purple-black, and from gold to lime green.

Avoid plums that have skin blemishes such as cracks, brown discolorations, or soft spots. A filmy, pale gray coating on the skin is natural and does not affect quality. Shriveled skin or leaking juice is a sign that they are too mature.

Store unripe plums loosely in a paper bag at room temperature until ripe. Ripe plums should be refrigerated in a plastic bag up to five days.

Popular Varieties: Santa Rosa, Black Beauty, Red Beauty, Friar. Two Main Categories: European (blue and purple, small, sweet, dried to make prunes) and Japanese (red and yellow, larger, less sweet, eaten fresh or canned).

Peak Season: August – September

Nutritional Content: 1 medium plum: 40 calories, 0.5 g fat, 9.5 g carbohydrates, 5 g sugar. Good source of vitamin C, source of vitamin A.

Did You Know...

- Pureed overripe plums frozen into ice cubes make excellent mineral water.
- Unlike many other fruits, a plum's sugar content continues to increase after it is picked.
- Plum trees produce fruit for twenty-five to thirty years.
- California plum production accounts for ninety percent of the plums eaten in the United States.
- During the California Gold Rush, the Pellier brothers became millionaires not from panning gold but from selling plums for several dollars each in fruit-deprived, scurvy-stricken mining towns.
- Plums originated thousands of years ago in the Caucasus Mountains nestled between Asia and Europe. Today there are almost two thousand varieties.

Pomegranate

Select pomegranates that have a bright red color, are blemish free, and heavy for their size. The skin should be shiny, taut, and smooth.

Avoid pomegranates that have dry, hard, wrinkled, or pale skins. Scarred areas or brown spots indicate old fruit. Also avoid fruit that have skin splits.

Store in a cool, dark place for several weeks or refrigerate up to two months.

Popular Varieties: Wonderful (most common, large, deep purple-red skin, crimson color flesh) and Sweet (slightly green with red blush, pink juice, very sweet).

Peak Season: September – February

Nutritional Content: 1 medium pomegranate: 100 calories, 0 g fat, 26 g carbohydrates, 21 g sugar. Good source of vitamin C, potassium, and fiber.

Did You Know...

- You must be careful when eating pomegranates; the rich scarlet juice will stain clothes, as well as fingers and nails.
- Pomegranates contain as many as eight hundred seeds and are exceptional appetizers. Their juice stimulates digestion.
- Many cultures use pomegranate seeds as a cure for nausea and the juice to remedy hangovers.
- Pomegranates have been cultivated since prehistoric times, and have appeared throughout history as symbols of fertility, royalty, faith, and abundance.
- Pomegranates were mentioned in the Old Testament several times under the name of Rimmon.
- Pomegranates derive their name from the Middle French words "pome garnete," which literally mean "seeded apple."

Prune

Select dried prunes that have bluish-black skins and are blemish free. They should be flexible, slightly soft, and purchased in sealed boxes or plastic.

Avoid dried prunes that are blemished or overly hard.

Store refrigerated in an airtight container up to six months. A prune's firmness and high sugar content prevents it from decaying and drying out.

Popular Varieties: Two Types: Fresh Prunes (sold as Italian plums or prune plums) and Dried Prunes (dried plums, most popular).

Peak Season: Available dried year-round

Nutritional Content: 5 dried prunes: 110 calories, 0 g fat, 25 g carbohydrates, 18 g sugar. Good source of vitamin A and fiber, source of vitamin C and potassium.

Did You Know...

- Substituting prune butter (pureed prunes) for butter when baking, cuts the calories by thirty percent and the fat content by seventy-five to ninety percent.
- Eating twelve prunes a day has been proven to lower cholesterol levels.
- Researchers say the nutritional content of prunes can help protect the body from certain types of cancer, heart disease, stroke, and Type II diabetes.
- Prunes contain double the antioxidants of any other fruit. Antioxidants are now believed to help slow the aging process in both the body and the brain.
- If all of the prunes grown in California in a single year were laid out in a line, it would circle the globe seventy times.
- Frenchman Louis Pellier, who came to California in 1848 in search of gold, first planted prunes in America in 1856. California now produces seventy percent of the world's prune supply.
- Because of a California labor shortage in 1905, farmers used monkeys to help harvest their prunes.

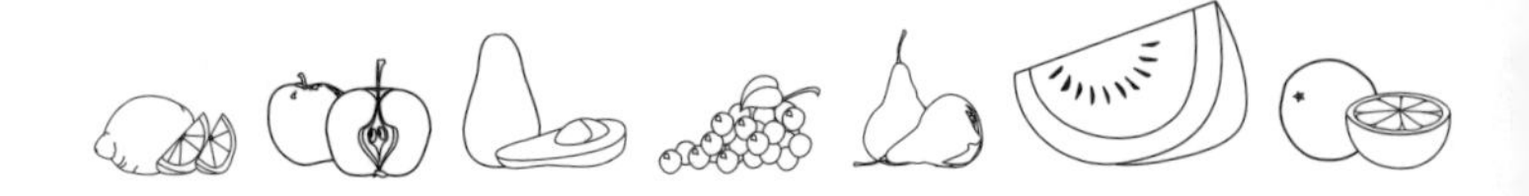

Raisin

Select raisins that are in tightly sealed packages. Exposure to air makes them dry and hard.

Avoid raisins that when shaken, rattle loosely in their package. This is a sign that they are dried out.

Store in a sealed plastic bag or container at room temperature for several months or refrigerate or freeze up to one year. Check occasionally for mold.

Popular Varieties: The most common grapes used for raisins are Thompson Seedless. They are sun-dried for several weeks, giving them a dark color and shriveled appearance. Golden raisins are also Thompson Seedless, but are dried with artificial heat leaving them plumper and moister than dark raisins. They are also treated with sulfur dioxide, which prevents darkening.

Peak Season: Available dried year-round

Nutritional Content: ½ cup raisins: 260 calories, 0 g fat, 62 g carbohydrates, 58 g sugar. Good source of fiber.

Did You Know...

- You can separate clumped raisins by placing them in a strainer and running hot water over them.
- An acre of grapes generates about four thousand pounds of raisins.
- Sugar makes up seventy-six percent of a raisin's content.
- Ninety-five percent of all raisins grown come from the San Joaquin Valley in California, which supplies most of the worlds market. Chile is also a major raisin producer.
- Drying raisins began more than three thousand years ago. The first dried raisins were most likely grapes left on the vine.
- Raisins were a cherished trade item in the Near East, and highly valued in ancient Rome.

Raspberry

Select raspberries that are firm, well formed, and have a bright, even color for their variety. They must be plump, dry, and fresh smelling. These berries are fragile because of their unique shape and hollow center. They should be used within twenty-four hours of purchase.

Avoid raspberries that are seeping juices and show signs of damage or mold. Attached hulls mean that the berries were picked too early and will be tart instead of sweet.

Store unwashed berries in the refrigerator up to three days. Remove moldy berries before storing to prevent decay from spreading to other berries. Wash berries right before using.

Popular Varieties: Red (most popular variety), Black (purple-black color, large), Golden (golden color, sweeter).

Peak Season: June – July

Nutritional Content: 1 cup raspberries: 50 calories, 0 g fat, 17 g carbohydrates, 12 g sugar. Good source of vitamin C and fiber, source of potassium.

Did You Know...

- An ounce of dried raspberry leaves boiled in a pint of water is a great remedy for relieving sore throats and healing canker sores.
- Raspberries are the most expensive berries because of their fragile nature, and the necessity of careful hand-picking.
- Raspberries are members of the rose family, and are native to North America and Asia.
- Raspberries have been cultivated since the time of Christ.
- Romans were most likely responsible for spreading raspberry cultivation throughout Europe because raspberry seeds were discovered in ancient Roman forts in Britain.

Rhubarb

Select rhubarb that has crisp, bright, fairly straight, firm, and smooth stalks. Reddish young shoots are preferred. Leaves should look fresh and blemish free. Stalk color should range from deep red to purple and fade to a green or white blush on both ends.

Avoid rhubarb stalks that are shriveled, soft, dull looking, or have brown or black ends.

Store refrigerated and tightly wrapped in a plastic bag up to three days. Remove leaves before refrigerating; wash just before using. Rhubarb perishes quickly. Rhubarb can be frozen and used for baking or making sauces.

Popular Varieties: Hothouse (sweet, tender, requires no peeling) and Field-Grown (strong flavor).

Peak Season: January – August

Nutritional Content: 1 cup diced rhubarb: 30 calories, 0 g fat, 4 g carbohydrates, 2 g sugar. Source of vitamin C.

Did You Know...

- Rhubarb leaves contain a poisonous acid and should never be eaten. They should be properly discarded out of the reach of children and pets.
- Rhubarb is an excellent digestive aid. It stimulates the taste buds, and cleanses the mouth. An ounce of rhubarb juice in the stomach increases the flow of gastric juice, which helps in processing stomach contents.
- Three teaspoons of rhubarb roots simmered for fifteen minutes, set aside overnight, and then strained, can be used to lighten hair to a more golden color.
- Rhubarb was used for medicinal purposes as far back as 2,700 B.C. in China. It made its way to Europe about three hundred years ago and is believed to have been brought to the New World by a Maine farmer in about 1800.
- Rhubarb is a cool season crop, and grows best in climates that have freezing winter temperatures. If rhubarb plants do not freeze, they likely will not produce the following season. To get rhubarb to grow in warm climates, the root generally needs to be removed from the ground and placed in the freezer for several months during the hottest part of the year.

Strawberry

Select strawberries that have perky green caps and a natural sheen. They should be firm, plump, well shaped, and brilliant red in color. A potent strawberry fragrance is a must. Medium size strawberries tend to have more flavor than larger ones. Check boxes for dampness or stains; this may indicate inferior fruit lies beneath the top layer.

Avoid strawberries that are shriveled, moldy, or without their green caps attached. Lack of fragrance, and any white at the tops, indicate they were immature when picked and may not fully ripen.

Store refrigerated in a large container until you are ready to eat them. Wash first, then hull just before eating. Washing or storing strawberries at room temperature greatly reduces their shelf life. For the richest flavor, strawberries are best served at room temperature. They are highly perishable; so eat them as soon as possible. They can be frozen in a heavy plastic bag up to a year.

Popular Varieties: Aromas (bright red sheen, large, firm, good flavor, fairly new variety), Camarosa (full red color, large, firm, good sheen, flavorful), Chandler (red color, large, flavorful), Diamante (bright red sheen, firm, large, flavorful, fairly new variety). There are two basic categories of strawberries: June bearing and ever bearing.

Peak Season: April – June (June-bearing varieties), early summer until frost (ever-bearing varieties)

Nutritional Content: 8 medium strawberries: 45 calories, 0 g fat, 12 g carbohydrates, 8 g sugar. Great source of vitamin C and fiber, source of potassium.

Did You Know...

- Eight strawberries contain more vitamin C than an orange.
- Strawberries are members of the rose family, and have approximately two hundred tiny seeds in one berry.
- If a single season of California's yearly harvest of strawberries were laid out in a line, it would circle the globe fifteen times.
- Strawberries were grown in Italy as far back as 234 B.C.
- Ancient Greeks were forbidden to eat strawberries because it was taboo to consume red foods.
- Romans cherished their medicinal qualities, using them to treat indigestion and toothaches.
- The largest strawberry ever grown was the size of an apple.

Tangelo

Select tangelos that are firm to slightly soft, heavy for their size, and smooth skinned with no deep groves. They are larger than tangerines and have a pebbly rind. Depending on the variety, color can range from light to deep orange, to reddish-orange.

Avoid tangelos that are dull or faded, have a rough or bumpy skin, or soft spots.

Store at room temperature for a week or so away from direct sunlight or heat, or refrigerate in a plastic bag up to two weeks.

Popular Varieties: Minneola or Honeybell (deep orange to red-orange color, large, bell shaped, smooth to pebbly peel) and Orlando (light to deep orange, medium to large, oval to round, pebbly peel, very juicy).

Peak Season: December – March

Nutritional Content: 1 medium tangelo: 45 calories, 1 g fat, 16 g carbohydrates, 8 g sugar. High source of vitamin C, good source of fiber.

Did You Know...

- Tangelos are a cross between a tangerine and a pomelo, a less frequent term for grapefruit. They are a type of Mandarin.
- One specific type of tangelo, a cross between a tangerine, a grapefruit, and an orange, discovered growing wild in Jamaica, is called "Ugli" because of its unsightly appearance.
- Although tangelos aren't as popular as other citrus varieties, they are gaining in popularity. Production now stands at eighty thousand tons a year and is increasing. They are grown in Florida, California, Arizona, and Texas.

Tangerine

Select tangerines that have a glossy, pebbly-textured skin. They should be heavy for their size, indicating juiciness. Skin should be loose fitting with a puffy appearance. Depending on the variety, color can range from pale yellow with a tint of green to deep or reddish-orange.

Avoid tangerines that have soft or brown discolored spots, or are dull or faded.

Store at room temperature for a week or so away from direct sunlight and heat, or refrigerate in a plastic bag up to two weeks.

Popular Varieties: Robinson (one-quarter grapefruit, cross between a clementine and an Orlando tangelo, large, vibrant orange-red flesh), Dancy (small to medium size, deep orange to red tinted skin, sweet, yet spicy), Honey (pale yellow skin with some greening, very fragrant, sweet, juicy flavor).

Peak Season: September – April

Nutritional Content: 1 medium tangerine: 50 calories, 0.5 g fat, 15 g carbohydrates, 12 g sugar. Good source of vitamin C, source of fiber and potassium.

Did You Know...

- Tangerines are the most popular and favorite citrus specialty. They are famous for their "zipper skins," which make them easy to peel.
- Tangerines were first grown in the Orient, where they were considered a rare luxury.
- The Imperial Japanese court had such a huge craving for tangerines that it became difficult to keep them in ready supply.
- Tangerines were once so rare, even their peels were cooked with sugar to make a sweet paste.
- Tangerines were named after the Moroccan city of Tangier after cultivation began along the North African coast.

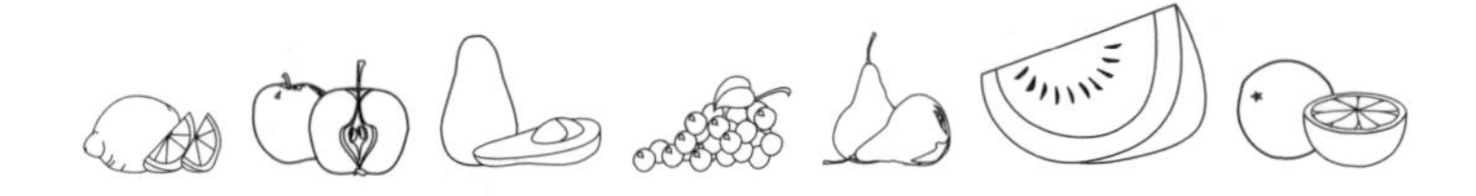

Tomato

Select tomatoes that are richly colored and well shaped for their variety. They should be firm, smooth skinned, and heavy for their size. If planning for later use, select tomatoes that are slightly yellow or orange.

Avoid tomatoes that have blotchy, brown, or green areas, broken or wrinkled skins, are bruised, or have mold around their stems.

Store ripe tomatoes stem side down at room temperature and use within a few days. Never refrigerate tomatoes. Cold temperatures make the flesh pulpy and diminish the flavor. Unlike many fruits, tomatoes will continue to ripen at room temperature after they are picked. Vine-ripened tomatoes have the best flavor.

Popular Varieties: Basic categories: Cherry (red or yellow, round, bite size, served in salads), Pear (red or yellow, small, pear shaped, sweet flavor), Plum/Roma/Italian (red or yellow, small, egg shaped, thick walled, rich flavor), Round/Slicing (bright red, most common, large, rounded, includes globe types and the flatter beefsteak variety).

Peak Season: June – September

Nutritional Content: 1 medium tomato: 35 calories, 1 g fat, 7 g carbohydrates, 4 g sugar. Good source of vitamin C and vitamin A.

Did You Know...

- By slicing tomatoes vertically from stem end to blossom end, rather than horizontally, it helps them retain both their shape and juice.
- Tomatoes are actually berries and are thought to have originated in the Andes Mountains of South America.
- Tomatoes were popularized in the United States by Creoles in New Orleans, who used them in their popular gumbo and jambalaya recipes.
- Up to one hundred fifty years ago, in both Europe and the United States, eating a tomato was thought to be fatal. That changed at noon on September 26, 1820, when Colonel Robert G. Johnson ate an entire case of tomatoes on the courthouse steps in Salem, Massachusetts, and lived to tell about it.
- Even though tomatoes are fruits, in 1893 the U.S. Supreme Court ruled that they must be considered a vegetable. This ruling was necessary because fruits and vegetables were subject to different import duties, and tomatoes were commonly eaten as a vegetable.

Watermelon

Select watermelons that are evenly colored, have a dull rind, and a pale or creamy yellow underside. Look for symmetry, either round, oblong, or oval depending on the variety. They must be firm, heavy in size, and yield to pressure. It is extremely difficult to tell if a watermelon is ripe by just looking; it must be examined. Look for the spot where the melon rested on the ground; a yellow-white spot suggests ripeness and a white or pale green spot indicates immaturity. Scratch the surface of the rind with your thumbnail. If the outer layer slips back with little resistance, showing a green-white color under the rind, the watermelon is ripe. Slap the side of the watermelon; if ripe, it will resonate with a hollow thump.

Avoid melons that are white or pale green, non-symmetrical, have gashes, soft spots, or other obvious rind blemishes. Also avoid those with a high-pitched tone or a dead, thud-like sound. When buying a cut melon, look at the color of the seeds, white seeds usually indicate that the melon was picked too early. Watermelons will not continue to ripen after they are picked.

Store refrigerated and consume within a week for optimal flavor. Watermelon can be stored uncut in the refrigerator for two or three weeks. Cut pieces should be tightly wrapped, refrigerated, and used in a couple of days. If the melon is too large for the refrigerator, keep it in a cool, dark place for a few days and use as soon as possible.

Popular Varieties: Four types: Picnic (red or yellow flesh, round or oblong, light to dark green rind with or without stripes), Ice Box (red or yellow flesh, round, dark or light green rind), Seedless (red or yellow flesh, oval to round, light green rind with dark green stripes), Yellow Flesh (yellow to bright orange flesh, oblong to long, light green rind with spotted stripes). There are two hundred varieties of watermelon.

Peak Season: May – August

Nutritional Content: 2 cups diced watermelon: 80 calories, 0 fat grams, 27 g carbohydrates, 25 g sugar. Good source of vitamin C, vitamin A, and potassium.

Did You Know…

- Watermelons, considered one of America's favorite fruits, are actually vegetables, cousins to cucumbers.
- Watermelons are grown in forty-four states, making the United States fourth in world production; China is first.
- Watermelons are ninety-two percent water.
- Russians make beer from watermelon juice.
- According to historians, watermelons have been cultivated for more than four thousand years. They were first grown in the middle of the Kalahari Desert and were a source of water for thirsty traders who began to sell the seeds in cities along ancient Mediterranean trade routes.

Resource List

Apricot Producers of California. www.apricotproducers.com

B.C. Tree Fruits Limited. www.bctree.com

California Apricot Council. www.califapricot.com/index.html

California Avocado Commission. www.avoinfo.com

California Cherry Advisory Board. www.calcherry.com

California Fig Advisory Board. www.californiafigs.com

California Kiwi Fruit Commission. www.kiwifruit.org

California Pear Advisory Board. www.calpear.com

California Prune Board. www.prunes.org

California Rare Fruit Growers: Fruit Facts. www.crfg.org/pubs/frtfacts.html

California Strawberry Commission. www.calstrawberry.com

California Table Grape Commission. www.tablegrape.com/rev99/index.htm

Chilean Fresh Fruit Association. www.cffa.org

Compucook. www.level6.com/Compucook/foodinfo/FICdata

Cranberry Creations. www.cranberrycreations.com

Cranberry Health and Nutrition. www.simmons.edu/~wry/use.html

Digitalseed: Fruits and Nuts. www.digitalseed.com/gardener/fruit

Dole 5 A Day. www.dole5aday.com

Driscoll's: The Finest Berries in the World. www.driscolls.com

Florida Department of Agriculture: Florida Commodities. www.fl-ag.com/commodities/index.htm

Florida Department of Citrus. www.floridajuice.com

Fresh Del Monte Produce Inc. www.freshdelmonte.com/produce_index.cfm

Fresh Point: Fresh Produce Information. www.freshpoint.com

Global Agribusiness Information Network: Postharvest Handling of *Breadfruit*. www.fintrac.com/gain/guides/ph/postbrdf.html

Grieve, M. *A Modern Herbal*. www.botanical.com/botanical/mgmh/comindx.html

Harris, Marilyn. *Common Tropical Fruit*. www.aloha.com/~ritt

Herbst, Sharon Tyler. *The Food Lover's Tiptionary*. New York: Hearst Books, 1994.

Hurst's Berry Farm. www.hursts-berry.com

Jubilee Foods: Nutrition Facts for Raw Fruits. www.jubileefoods.com

Keany Produce Archives. www.keanyproduce.com/archives/archives.htm

Larse Farms: Strawberries. www.sweetdarling.com

Melissas: Product Information. www.melissas.com

Mid City Nursery. www.midcitynursery.com

National Watermelon Promotion Board. www.watermelon.org

Ohio State University Extension Fact Sheet: Selecting, Storing, and Serving Ohio Melons. www.ag.ohio-state.edu/~ohioline/hyg-fact/5000/5523.html

Oregon Raspberry and Blackberry Commission. www.oregon-berries.com

Produce Oasis: Main Alpha Page. www.produceoasis.com/Alpha_Folder/Alpha.html

Rieger, Mark. *Mark's Fruit Crops.* www.uga.edu/hortcrop/rieger/#Crops

Rinzler, Carol Ann. *The New Complete Book of Food: A Nutritional, Medical, and Culinary Guide.* New York: Checkmark Books, 1999.

Rombauer, Irma S. and Marion Rombauer Becker. *Joy of Cooking.* New York: Bobbs-Merrill Company, 1984.

Sea View Brand: Medjool Dates. www.seaviewsales.com/index.html

Sharvy, Ben. *Edible Landscaping and Gardening.* www.efn.org/~bsharvy/edible.html#ediblesites

Solutions Series. www.ag.uiuc.edu/~robsond/solutions/horticulture/docs

Sunkist Growers. www.sunkist.com.

Sunmaid Growers of California. www.sunmaid.com

Texas Blueberries. www.tx-marketeers.com/TxBlueberries

The California Avocado Commission. www.avoinfo.com

The Exotic Fruits of Malaysia. agrolink.moa.my/comoditi/fruits.html

The Fruit Pages: Everything You Want to Know About Fruit. www.thefruitpages.com

The Mid-Atlantic Regional Fruit Loop. www.caf.wvu.edu/kearneysville/fruitloop.html

The Natural Food Hub: Fruit Western People Commonly or Occasionally Eat. www.naturalhub.com/natural_food_guide_fruit_common.htm

The Original Strawberry Facts Page. www.jamm.com/strawberry/facts.html

Tropico: Industrial Fruit Division. www.tropico2000.com

University of California: Fruit & Nut Research and Information Center. fruitsandnuts.ucdavis.edu

University of Illinois Extension: Apples & More. www.urbanext.uiuc.edu/apples/facts.html

USDA Agriculture Research Service: Nutrient Data Laboratory. www.nal.usda.gov/fnic/cgi-bin/nut_search.pl

Valley Fig Growers. www.valleyfig.com

Viard, Michael. *Fruits and Vegetables of The World*. Paris: Longmeadow Press, 1995.

Washington State Apple Commission. www.bestapples.com

Washington State Fruit Commission. www.nwcherries.com

Wegmans: Fresh Produce. www.wegmans.com/kitchen/ingredients/produce/index.html

WholeHealthMD.com. www.wholehealthmd.com

Wonderful World of Guavas Homepage. www.ocf.berkeley.edu/~montymex/guavaintro.html

About the Author

T.M. Gorman's journey from fruit novice to fruit expert began when she grew tired of bringing home dry, mealy, tasteless fruit from the market. An avid reader, she soon realized that a user-friendly, quick-reference guide to selecting fruit was not readily available to consumers. After much research and practice, she finally learned the secrets of quality fruit selection. In *Fruit–The Ripe Pick*, she shares her newfound knowledge with all others in search of delicious fresh fruit. Gorman is a native of Hawaii, and resides on the island of Oahu with her husband Richard and their dog Koko.

Notes

Notes

Notes

If you enjoyed this book and would like to pass one on to someone else, please check with your local bookstore, online bookseller, or use this form.

Please Send Me:	Amount
___ Copies of *Fruit—The Ripe Pick* @ $9.95 Each	________
___ Shipping & Handling: $2.55 Per Book	________
Total Amount:	________

Payment: ☐ Check ☐ Credit Card ☐ Money Order
Make check or money order payable to: Vine Publishing
Visa ☐ MasterCard ☐

Card #: ____________________________ Exp. Date ____ /____

Name On Card: ____________________________________

Signature: ____

Shipping Information:
Name: ____

Address: ____

City/State/Zip: ____________________________________

Phone: (808) 235-1730
Fax: (808) 235-1735
Mail: Vine Publishing
PO Box 17912
Honolulu, HI 96817-9998
e-mail: vine@hawaii.rr.com

Reader Comments: ________________

If you enjoyed this book and would like to pass one on to someone else, please check with your local bookstore, online bookseller, or use this form.

Please Send Me: Amount

___ Copies of *Fruit—The Ripe Pick* @ $9.95 Each ________

___ Shipping & Handling: $2.55 Per Book ________

Total Amount: ________

Payment: ☐ Check ☐ Credit Card ☐ Money Order
Make check or money order payable to: Vine Publishing
Visa ☐ MasterCard ☐

Card #: ______________________________ Exp. Date ____ /____

Name On Card: ______________________________

Signature: ______________________________

Shipping Information:
Name: ______________________________

Address: ______________________________

City/State/Zip: ______________________________

Phone: (808) 235-1730
Fax: (808) 235-1735
Mail: Vine Publishing
PO Box 17912
Honolulu, HI 96817-9998
e-mail: vine@hawaii.rr.com

Reader Comments: ____________